POEMS & PLAYS ABOUT WEATHER AND CLIMATE

Brod Bagert's HeART of Science
www.brodbagertsheartofscience.com
877.968.7522

Published by Living Road Press
San Antonio, TX

ISBN 978-1-7321515-4-3

Available at www.brodbagertsheartofscience.com
877.968.7522

Dedication

To my wonderful daughter-in-law, Celeste Lofton-Bagert, B.B.
To my dear niece, Arina Kolosova, N.K.

Acknowledgements

A big thanks to Eric M. Schlegel, Ph.D., Vaughan Family Professor, Physic & Astronomy, Astrophysicist at the University of Texas at San Antonio for making certain all concepts are accurately represented.

Introduction—Weather or climate? Climate or weather?

Not quite the same but so close together.
Climate or weather? Weather or climate?
The difference is simply the way that you time it.
from "Whether It's Weather"

This book is about weather and climate, which some of you may already know are not the same thing, even though they are very closely related. So let's start with weather.

The best thing about weather science is that you experience weather every day, and understanding something you experience every day makes each day more interesting. This book is packed with all the information you need to understand climate and weather. When you see a cloud in the sky, it's not just any cloud, it's a particular kind of cloud, it's got a name, and you'll learn all about it. You'll understand why wind makes you feel colder in winter and why humidity makes you feel hotter in summer. The awesome spectacle of a lightening storm will be even more awesome when you understand what's happening inside those clouds to make all that electricity.

Knowing the basics of weather science is just plain fun, but climate is another thing entirely. There is no longer any question that Earth's climate is changing, but there are a bunch of questions about what we should be doing about it, real questions that require hard decisions. Most of you will not become climate-change scientists, but all of you will be part of making those decisions, which is why it's important that you understand the complexities of weather science. How important? Really-really-really important.

Here is something to think about. **Hurricanes are weather** and they're a big deal. My first memory of a hurricane goes way back to the year I started first grade. School was cancelled, there was a feeling of magic in the air, big oak tree branches shook in the wind—it was very exciting, and I remember being a little disappointed when the hurricane changed course and missed us entirely.

But **the gradual increase in average global temperature is climate**, and it's a very big deal, bigger than a thousand hurricanes, and it's not the kind of thing that can just change course and go away. It has already begun, and you are already a part of deciding what we do about it. It's important stuff, and I'm hoping this book will make learning about it, talking about it, living it, and predicting it, electrifying.

TABLE OF CONTENTS

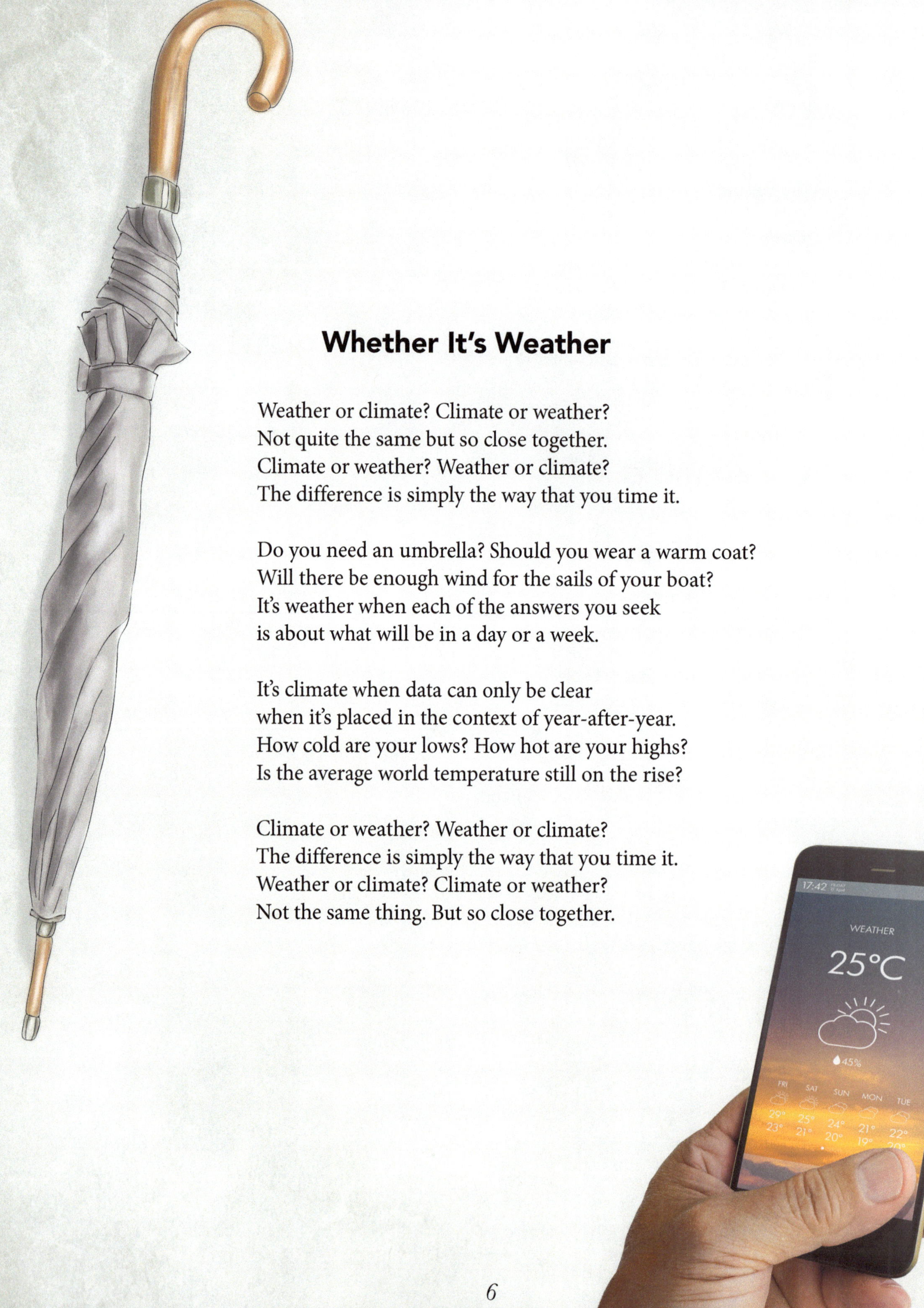

Whether It's Weather

Weather or climate? Climate or weather?
Not quite the same but so close together.
Climate or weather? Weather or climate?
The difference is simply the way that you time it.

Do you need an umbrella? Should you wear a warm coat?
Will there be enough wind for the sails of your boat?
It's weather when each of the answers you seek
is about what will be in a day or a week.

It's climate when data can only be clear
when it's placed in the context of year-after-year.
How cold are your lows? How hot are your highs?
Is the average world temperature still on the rise?

Climate or weather? Weather or climate?
The difference is simply the way that you time it.
Weather or climate? Climate or weather?
Not the same thing. But so close together.

Metling Ice Formations in Antarctica

The difference between weather and climate is very simple, but it's very, very important.

Weather tells us what the atmosphere is doing at a particular place and time. **Climate** tells us what the atmosphere does over a long period of time. So the question What's the temperature today in Buford, Wyoming? is a question about weather. But the question What's the average annual temperature in Buford, Wyoming? is a question about climate.

Chariot of Fire
A Statement from the Sun

I am the Sun.
The author of this book asked that I give him an interview.
He seems to think that no one can understand Earth's weather
without first understanding ME.
I told him I was much too busy for his foolishness
and commanded him to leave my presence at once,
but he is a very insistent fellow,
so, in lieu of a personal interview,
I agreed to give him a statement.
Therefore:

I, THE SUN, MAKE THE FOLLOWING STATEMENT OF MY OWN FREE WILL AND DO HEREBY CERTIFY THAT THE CONTENTS HEREIN ARE TRUE AND CORRECT.

> The first thing you should know is that your weather,
> all of your weather, begins with ME.
> In the simplest terms,
> I am a ball of hydrogen transforming itself into helium
> by a process known as nuclear fusion,
> a byproduct of which is the emission of energy
> in the form of electromagnetic radiation,
> mostly the center of that spectrum which consists of
> infrared rays, visible light, and ultraviolet rays,
> though during my occasional flares I emit deadly gamma rays as well,
> all of which radiates from my surface in all directions
> arriving at the edge of your atmosphere in approximately eight minutes.
> (Your atmosphere. YOUR atmosphere.
> You Earthlings should take better care of your atmosphere.
> It is your protective shield—
> all that ozone and carbon dioxide and those floating masses of water vapor.
> Without them I would burn you to a crisp.
> And too much of that carbon dioxide and I would bake you in an oven.)
> Now, though it pains me to say it, your pesky author is correct;
> to understand Earth's weather you must first understand ME.
> The angle of my rays as they enter your atmosphere bring the seasons.
> I expand the gases of your atmosphere and make the wind.
> I transform water from liquid to gas as the engine of your water cycle.
> I drive the currents of your oceans.
> I...am...the BOSS.

IN DECLARATION THEREOF I DO HEREBY AFFIX MY SIGNATURE:

The Sun
AKA - CHARIOT OF FIRE

When it comes to climate, the Sun really is the boss. The angle of the Sun's rays entering the atmosphere creates vast cycles of moving air that determine local weather conditions all over the world. If you decide to delve deeper into meteorology you'll learn about things like Ferrel, Hadley, and Polar wind cells; Westerlies, southeasterly, and northeasterly trade winds; polar and subtropical jet streams.

The interaction of these cycles with ocean currents also plays an important role. The gulf stream is an ocean current we hear a lot about, but it's only one of 16 major ocean currents around the world. Variations in each of these currents have a powerful impact on weather conditions.

The famous El Niño (the boy) is a recurring climate cycle in the Pacific Ocean. It happens every few years when there is a slight eastward shift of warm water in the Pacific as it moves along the equator toward South America. In simplest terms, El Niño means lots of lots of rain in California, fewer hurricanes in the north Atlantic, and warmer winters in North America.

La Niña (the girl) is another recurring climate cycle in the Pacific and it has the opposite effect. It happens when unusually strong trade winds push warm waters in the Pacific toward the west causing cold water to rise from the depth of the ocean. Again, in simplest terms, La Niña means colder winters in North America and more hurricanes in the eastern Pacific.

The actual effects of El Niño and La Niña are very complex and impact weather conditions around the world. The important thing to understand is that shifts in global wind patterns can alter the temperatures of ocean currents and cause dramatic changes in weather worldwide.

Dangerous Planet
Human Courage

Tommy Rain: 7 year-old boy, harmlessly mischievous.
I am a rainstorm. I make you wet.
I'm 'bout as bad as the weather can get.
When you see my bright flashes and thunderous roars
you should shut all the windows and lock all the doors.

Sandstorm Rodriguez: 10 year-old girl, proud and loud.
Did you hear what he said, that poor little chump?
On the road of bad weather, he's barely a bump.
But sandstorms, like me, we lay down the law.
Our wind-driven sand can scrape paint off your car.
Out of nowhere we rise on a hot desert day.
When you see one of us just get out of the way.

Latoya Blizzard: 12 year-old girl, fast-talking, bossy, in-your-face.
Yes, sandstorms are naughty, but surely you know
that we blizzards can bury your houses in snow.
We can bury the highways in gray icy-cold
until all transportation is out of control.
When the weatherman says, "A severe winter-storm,"
you should find a good book and just try to stay warm.

Freddy Tornado: 14 year-old gangster, slow, insincere, threatening.
That's right. They're all bad. 'Bout as bad as can be.
So don't even worry 'bout little-ole me.
When we little tornadoes drop out of the sky
we can toss eighteen-wheelers a hundred feet high,
but we give little warning, and that's why you should see
that it's senseless to worry 'bout little-ole me.

Hurricane: Wicked Witch.

Those bad little storms can be messy, it's true,
but nothing compares to the damage I do.
There's nowhere to run and there's nowhere to hide
when I'm huffing and puffing three hundred miles wide.
When the whole world around you is going insane
with the scream of my wind and the crash of my rain,
you will know it was silly to work and to plan
as I bulldoze the ocean up over the land.

All together:

When humans meet us, they meet fear face-to-face,
and they know Mother Earth is a dangerous place.
Yet with all of that danger, and all of that fear,
you humans keep facing us, year after year.
So in case you don't know it, it's time that you see
human courage is tougher than we'll ever be.

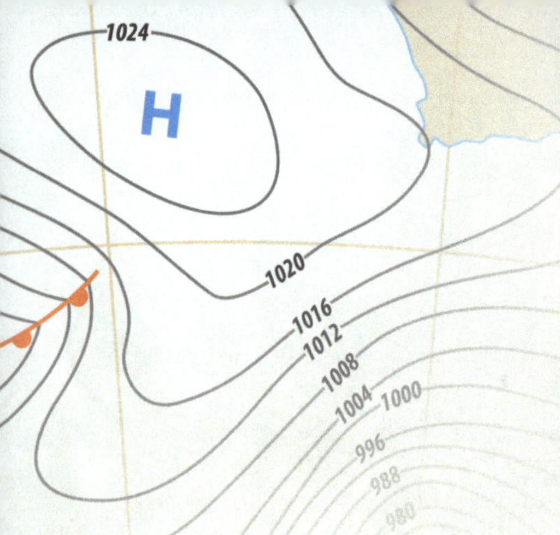

Need to Know

Forecast-forecast! What will be?
Tell me what's in store for me.
How much wind and rain and snow?
Tell me now, I NEED TO KNOW!

Lightning! Thunder! Oh the dread!
Big fat rain drops on my head.

Dust storm! Dust storm! Hot and dry!
Dust so thick it hides the sky.

Blizzard! Blizzard! Ice and snow!
Stay inside, nowhere to go.

Wild tornado! Hear that roar!
In the cellar! Lock the door!

HUR—RI—CANE! Holy cow!
EVACUATE! DO IT NOW!

What's the forecast? What will be?
Tell me what's in store for me.
How much wind and rain and snow?
Tell me now, I NEED TO KNOW!

Meteorologists are scientists who study the atmosphere to understand what causes changes in the weather and how to predict it. Meteorologists have learned to forecast some pretty important things, like the probability of tornadoes and the path and intensity of hurricanes. These kinds of predictions often give us the time we need to protect ourselves from the danger of extreme weather.

Measure the Weather

Mr. Thermometer, tell-me, tell-me true.
What exactly, what exactly do you really do?
Fahrenheit or centigrade,
with me, my friend, you've got it made.
I'm a pleasure, I'm a treasure,
temperature is what I measure.

Please Señor Hygrometer, tell-me, tell-me true.
What exactly, what exactly do you really do?
Forest damp or desert dry,
I will be your favorite guy.
I'm a pleasure, I'm a treasure,
humidity is what I measure.

Mademoiselle Barometer, tell me, tell me true.
What exactly, what exactly do you really do?
Pressure changes, high or low,
tell you what you need to know.
I'm a pleasure, I'm a treasure,
I gage atmospheric pressure.

Professor Anemometer, tell me, tell me true.
What exactly, what exactly do you really do?
Wind speed? Let me see.
How fast did that wind blow by me?
I'm a pleasure, I'm a treasure,
moving air is what I measure.

Each of us an *-ometer*, so what we say is true,
'cause measure-measure-measuring is what we love to do.
It's a joy! It's a treasure!
Measure-measure-measure-measure!

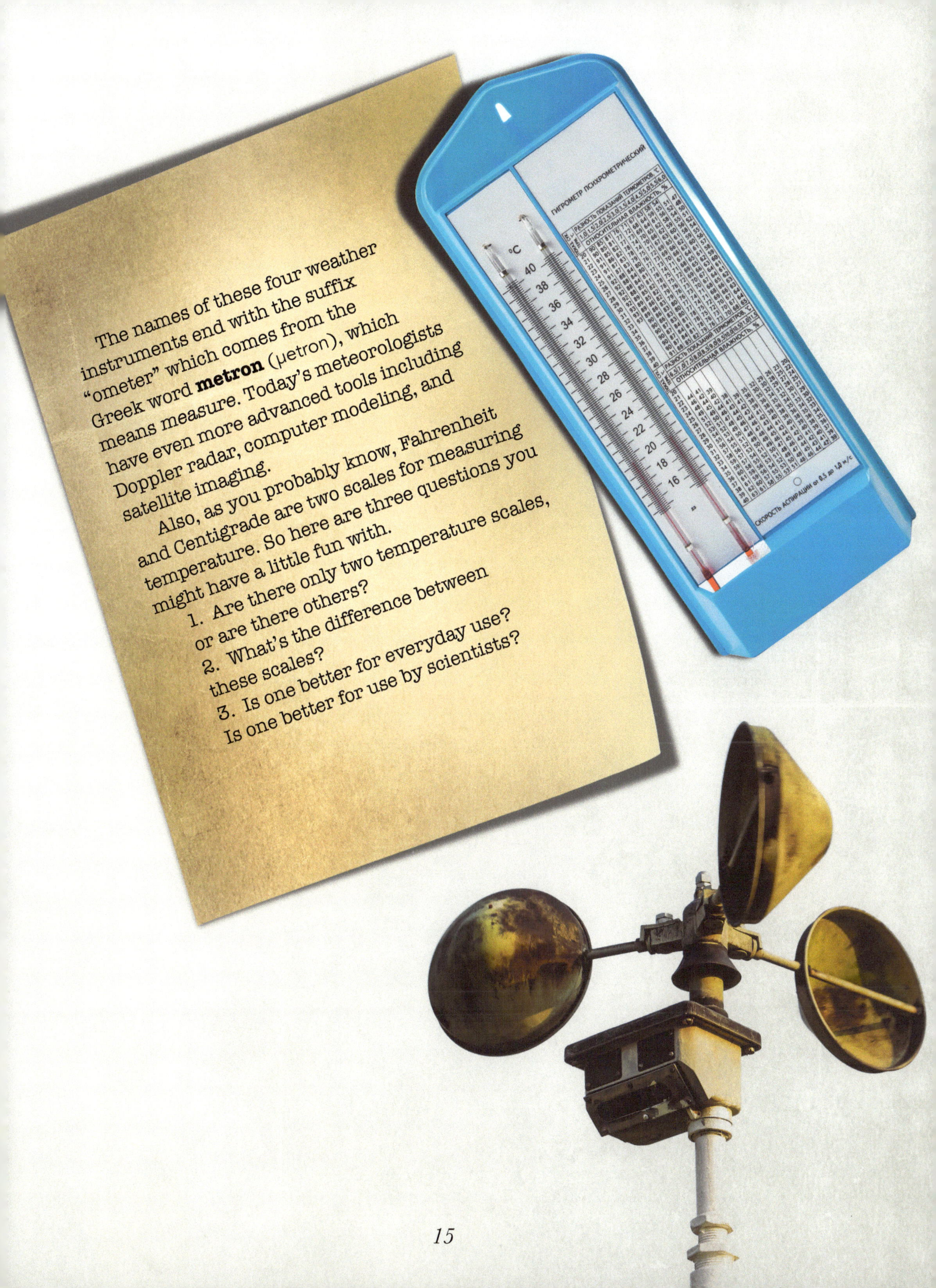

The names of these four weather instruments end with the suffix "ometer" which comes from the Greek word **metron** (μετρον), which means measure. Today's meteorologists have even more advanced tools including Doppler radar, computer modeling, and satellite imaging.

Also, as you probably know, Fahrenheit and Centigrade are two scales for measuring temperature. So here are three questions you might have a little fun with.

1. Are there only two temperature scales, or are there others?
2. What's the difference between these scales?
3. Is one better for everyday use? Is one better for use by scientists?

Taking Turns

Summer sunshine, summer heat,
summer sandals on my feet.
Lots of time for me to play
summer please don't go away.

Autumn, autumn—turning cool.
Time for me to go to school.
Something tingles in the air…
I see pumpkins everywhere!

Winter, winter—ice and snow,
bundle up from head to toe.
Snow at sunset, snow at dawn,
soon the snow will all be gone.

Springtime! Springtime! Spring is here!
Springtime flowers make me cheer!
Springtime showers, warm and clean!
Everything I see is green!

Every year, it's only fair,
all the seasons have to share.
All the seasons have to learn
that every season takes a turn.

Earth's seasons are caused by the angle at which the Sun's rays enter Earth's atmosphere. That angle changes throughout the year as Earth moves in its orbit around the Sun: steeper in summer; more slanted in winter.

Now, I'm about to say what causes that change. It's really simple, but it sounds complicated. You don't have to understand it right away, just say the words out loud to make them familiar in your ears. Here it is: The Earth rotates on its axis at an angle of 23.5 degrees relative to its orbital plane. Astronomers call this **axial tilt**.

So the next time someone asks you what causes the seasons, smile casually and say, "We have seasons because Earth rotates on its axis at an angle of 23.5 degrees relative to its orbital plane around the Sun." Boom! You're a genius! (More about this in the next poem.)

The Angle of Our Seasons

Earth rotates at an angle—
twenty-three point five degrees.
A tilt that makes our summers hot
and makes our winters freeze.

When half of Earth tilts toward the Sun
the air begins to heat,
and that's the time we grow our food
like lima beans and wheat.

But when Earth tilts the other way
as every year it will,
the icy wind comes rolling in
and life begins to still.

And in between when leaves turn green
Hip-hip-hurray! It's spring!
Or when the leaves turn red and gold,
and autumn colors sing.

Without that tiny axial tilt
our lives would rearrange—
Always hot! Or always cold!
The seasons couldn't change!

So an angle brings the seasons,
every year with grace and ease.
The cycle's set, so don't forget—
Twenty-three…point five…degrees.

EARTH'S SEASONS
in the Northern Hemisphere

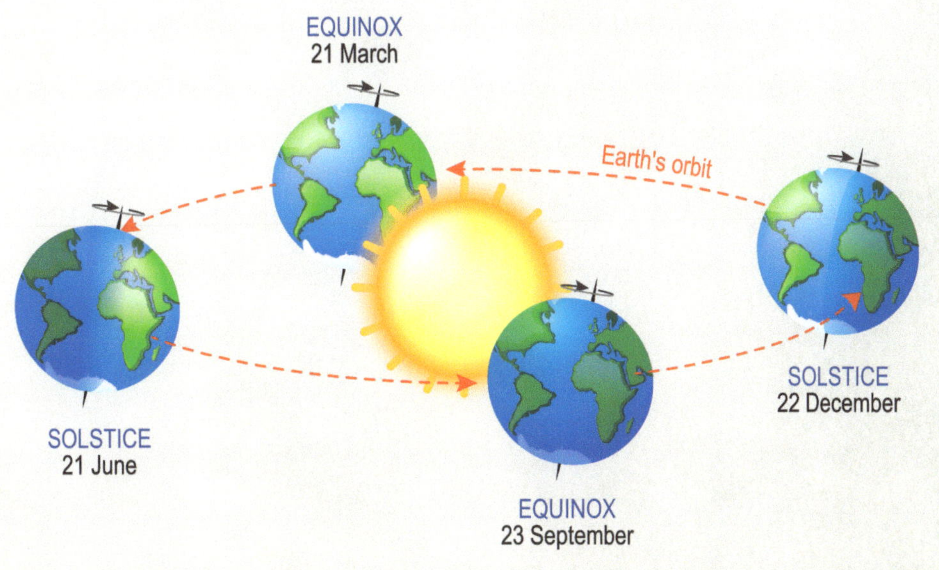

Get ready. This is one of my favorite things about seasons and the night sky. It took me a while to understand what I'm about to share with you, so don't be discouraged if you don't get it the first time through. We're going to start off with the seasons and end up with the constellations of the Zodiac. Here we go.

Every day the Sun moves across the sky, rising in the east and setting in the west. In summer, the Sun is high in the sky which makes it hot. In winter, the Sun is low in the sky which makes it cold. In spring and autumn, the Sun is in between its high and low points which makes the weather mild. Each of these seasons has a kind of milestone with an official name: **summer solstice** (late June, the Sun is at its highest), **autumnal equinox** (late September, the Sun is half way down), **winter solstice** (late December, the Sun is at its lowest), **vernal equinox** (late March, the Sun is half way back up).

Now the path of the Sun on the equinox days, when the Sun is right in the middle of its highest point in summer and lowest point in winter, that path has a name. It's called the **celestial equator**. And that path across the sky is exactly 23.5° below the summer high-point and 23.5° above the winter low-point. So between the summer high-point and the winter low-point there is an arching band of sky that is exactly 47° wide (23.5° + 23.5°) and that band also has a name. We call it the **zodiac**.

The zodiac is a place, it's a 47° band that arcs across the sky from east to west, and it's a very interesting place because that's where you have to look if you want to see the Sun and the Moon and all the planets—all of them, always inside the zodiac, never outside of it. And at night you'll see a series of constellations, 13 in all, that appear in that same band of sky, and we call them the constellations of the zodiac.

Astronomy and astrology are very different things. Astronomy is science and astrology is a little like fortune telling. But the zodiac is where the two share a bit of common ground.

Only One

Little snowflake, floating free,
It seems that you're a lot like me.

You're unique, oh-yes it's true,
there's never been another you.
So we're alike, it's plain to see,
because there's only one of me.

But there's one way we're not the same,
a difference I will now proclaim,
to me the most important thing—
I do not have to melt in spring.

Is it true that no two snowflakes are alike? The only way to know for sure would be to examine every snowflake that ever was, which is clearly impossible. But, based on the way snowflakes form, scientists think it's probably true, and here's why. A typical snowflake contains about 10 quintillion water molecules. Each one of those molecules freezes to form a crystal around an ice nucleus (a speck of dust) as it falls through the atmosphere toward the ground.

A snowflake grows as it falls, but how it grows depends on the speed at which it falls, along with the temperature and humidity of the atmosphere, all of which are constantly changing, and every tiny change affects the snowflakes final shape. That's a lot of molecules and a lot of changes, which is why it's probable that no two snowflakes ever end up with the exact same shape.

The Puddle Bandit

Can a puddle full of water
simply up and disappear?
I passed one on my way to school,
but now it isn't here.

And it isn't just today.
It happens all the time.
Somebody's stealing puddles
which has got to be a crime.

I could set a trap and catch him,
I could find a place to hide,
but the Sun is way too hot today,
I've got to get inside.

The speaker says that he can't hideout to catch the puddle bandit because the Sun is too hot, but the heat of the Sun is exactly what's causing the puddle to evaporate. It's a little joke that contains two big questions. What is evaporation? And how does the Sun evaporate water?

Evaporation is the term we use to describe the phase change that happens when matter changes from a liquid to a gas. When liquid water evaporates it turns into a gas we call **water vapor**. So how does the Sun evaporate water?

All matter is made up of atoms and molecules, and all those atoms and molecules are always in motion. In cold matter, the molecular motion is slow. As matter gets hotter and hotter, its molecular motion gets faster and faster. So water in a puddle is a bunch of H_2O molecules moving slowly enough to remain in a liquid state. As the Sun heats the puddle, the H_2O molecules move faster and faster until they get fast enough to take off into the air as water vapor.

From now on, you can think of heat as molecular motion. In another book in this series, we'll explain exactly what **molecular motion** means, but for now just think of it as molecules vibrating and swirling around at the same time.

CONDENSATION

EVAPORATION

No Vacation

Runoff and accumulation!
H_2O gets no vacation!
Yes-oh-yes, oh-yes I know!
Watch the water come and go.
Starting high and running down,
on the surface, underground—
a river winding like a snake,
a tiny pond, a giant lake.
Water loves to be in motion,
always headed to the ocean.

Yes-oh-yes, oh-yes I know!
Watch the water come and go.
What comes next? Evaporation!
H_2O gets no vacation!
When the Sun gets very hot
water doesn't have a shot.
In a lake or in a stream
sunshine turns it into steam.
And POOF! The water isn't there.
It floated up into the air.

Yes-oh-yes, oh-yes I know!
Watch the water come and go.
H_2O gets no vacation!
What comes next is condensation!
Here's a scientific rule—
warm stuff always wants to cool.
And cooling vapor never stops
condensing into tiny drops,
forming clouds up in the air,
floating-floating everywhere.

ACCUMULATION

Yes-oh-yes, oh-yes I know!
Watch the water come and go.
H_2O gets no vacation!
Here it comes! Precipitation!
Now the clouds are dark and gray.
It's going to be a rainy day.
A zillion droplets, maybe more.
It won't be long, it's going to pour.
Rain-rain! Go away!
Come again another day!

ACCUMULATION! EVAPORATION!
CONDENSATION! PRECIPITATION!
H_2O! H_2O!
H_2O! GETS NO VACATION!

At a temperature of 90°F in the desert you would feel a lot cooler than at 90°F in the swamp. This is because of a thing called the **heat index**, and evaporation is what makes it work.

The **heat index** uses two pieces of data: degrees of temperature and percent of **humidity**. Temperature tells us how warm the air is, **humidity** tells us how much water vapor it contains. The higher the humidity, the warmer the air feels to your human body.

Here's why. Sweat, which is mostly water, is one of the things your body uses to stay cool. As sweat evaporates from your skin, it removes heat from your body, and you feel cooler. But in high humidity, sweat evaporates much more slowly because the air is already full of water vapor. So in low humidity desert air, your sweat evaporates quickly and you feel a lot cooler than you would feel at the same temperature in high humidity swamp air.

High humidity, by slowing down the evaporation of sweat on your skin, makes the air feel warmer. **Heat index** is the way we express precisely how much warmer.

A Problem with the Water Cycle
A Disgusting Realization

My cute little puppy walks up to a tree,
he sniffs it up and down,
then he lifts his leg, looks directly ahead,
and pee-pees on the ground.

At first the pee just sits there in
the midday heat and glare,
then slowly it evaporates
and floats up the air.

Now vapor in the air I know,
will end up in a cloud.
I'm very good at science, and
it makes me very proud.

Clouds make rain, rain gathers in creeks,
and creeks flow toward a river.
(Now this is the really disgusting part
that makes my stomach quiver.)

I'm really reluctant to say it,
I hate to even think it,
but fresh water in a river?
Some human will eventually drink it.

Earth's water is almost four billion years old,
which is why it's seems likely to me,
that every drop we humans drink
was once in a puddle of pee.

This disgusting bit of trivia is probably true. One of my favorite books about science is *A Short History of Nearly Everything* by Bill Bryson. Here's a quote from a chapter called "The Mighty Atom":

"Atoms are fantastically durable. Because they are so long lived, atoms really get around. Every atom you possess has almost certainly passed through several stars and been part of millions of organisms on its way to becoming you. We are each so atomically numerous and so vigorously recycled at death that a significant number of our atoms—up to a billion for each of us, it has been suggested—probably once belonged to Shakespeare."

My Favorite Cloud

Some clouds are white and fluffy,
and some are dark and gray,
but every cloud gets started
in a very simple way.

First water turns to vapor—
summer, winter, fall, or spring.
But vapor is invisible,
your eyes can't see a thing.

It rises in the atmosphere
it's headed for the top,
but soon the vapor starts to cool
and makes a tiny drop.

Then drops bunch up together
in a kind of floating crowd,
and when you get enough of them,
that's when you see a cloud.

And when that cloud gets dark and gray
you need not guess or wonder,
pretty soon you'll see a flash
and then you'll hear the thunder.

I know I shouldn't be afraid,
I know it's in my head,
but thunder makes me want to run
and hide beneath my bed.

Those big-ole noisy thunder clouds,
they rattle all my fears.
That's why I like the clouds that never
thunder in my ears.

What causes the sound of thunder is the same thing that causes the boom of a firecracker—hot air under pressure expanding very rapidly.

In a firecracker, black gun powder is packed tightly inside a paper wrapper. When lit, the burning powder becomes a very hot gas that wants to expand outward putting pressure against the paper wrapping. When the pressure of the hot gas exceeds the pressure of the containing wrapper, the gas explodes outward and BOOM!

Much the same thing happens with lightning. A bolt of lightning heats the air around it to a peak temperature of over 48,000°F. It happens so quickly that the heated air is compressed by the cooler air around it, 10 to 100 times normal atmospheric pressure. When the pressure of the heated air exceeds the pressure of the containing cooler air around it, the heated air explodes outward and THUNDER!

contrails

sirrocumulus

Flying Water

Water! Water! Way up high!
Water flying in the sky!
Look up here! Up in the air!
We clouds are almost always there!
  I'm a Cirrus, way up high,
  icy crystals in the sky.
  Way up high and wispy white?
  Say it! Cirrus! That's right.

Look up here! Up in the air!
We clouds are almost always there!
  Cumulus! Yes! Yes-siree!
  I'm as fluffy as can be.
  Sometimes high. Sometimes low.
  I'm a cloud you need to know!

nimbus

stratocumulus

Here's an easy way to use the vocabulary of clouds. We start with the names of the three basic kinds of clouds: **cumulus**, starts low and puffs high into the air, **stratus**, low and flat, and **cirrus** high and wispy.

Now you can add the word **nimbus** to cumulus and stratus. Nimbus is a Latin word that means *storm*. So a **cumulonimbus** cloud is a big fluffy dark cloud full of water and ready to rain. And a **nimbostratus** cloud is a low, flat, dark cloud full of water and ready to rain. I have no idea why we put the nimbus after one and before the other, they just do.

By the way, it should come as no surprise to you that as a poet I love the sound of words. And of all the words I have ever heard, my very favorite is the word **cumulonimbus**. I love the feel of the vowels as they roll around in the middle of my mouth. It also has a wonderful rhythm consisting of two poetic feet: a dactyl and a spondee. So it sounds like this—DA-da-da DA-DA—which is exactly the rhythm at the end of every line in Homer's great epic poem, *The Odyssey*. Please forgive me, sometimes I can't help myself.

Look up here! Up in the air!
We clouds are almost always there!
I'm a Stratus, lower down.
Sometimes I can touch the ground.
That's when I'm fog, imagine that,
but high or low I'm always flat.

Look up here! Up in the air!
We clouds are almost always there!
Nimbus! Nimbus! Don't complain!
Nimbus means it's going to rain.
Rain or sleet or snow or ice,
nimbus weather's never nice.

Look up here! Up in the air!
We clouds are almost always there!
Water! Water! Way up high!
Water flying in the sky!

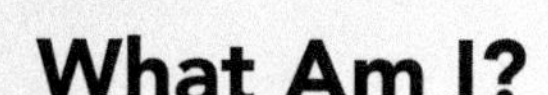

What Am I?

In dark of night I come and go.
If you're inside you'll never know,
'cause even if I'm all around
you'll never hear me make a sound.

My misty-murk on scary nights
can swallow up your city-lights,
and in that cloak of shrouded air
who knows what might be creeping there?

But wait, why do you back away.
Oh please, don't scream and run away.
And please-oh-please don't start to cry—
I'm just a cloud that's scared to fly.

Fog is a cloud that is so low it touches the surface of the ground. It you watch fog as it lifts, you will see that it's just an ordinary cloud.

A few years ago, I had a chance to experience something a little like this in Hawaii. I was driving an open Jeep up a winding road that lead to the top of Haleakala, a 10,000 foot high volcano that forms more than 75% of the Island of Maui. To my left and right, I could see the Pacific Ocean, but there was a layer of clouds overhead and I couldn't see the top of the mountain. As I drove higher and higher, the layer of clouds got closer and closer, then suddenly I was in it. What a few minutes before had been a cloud overhead was now fog all around me, but the best was still to come. I continued to drive, higher and higher as before, when just as suddenly I broke out of the cloud, and what a view! The peak of Haleakala above me, the clouds below, and in every direction, the blue magic of endless ocean.

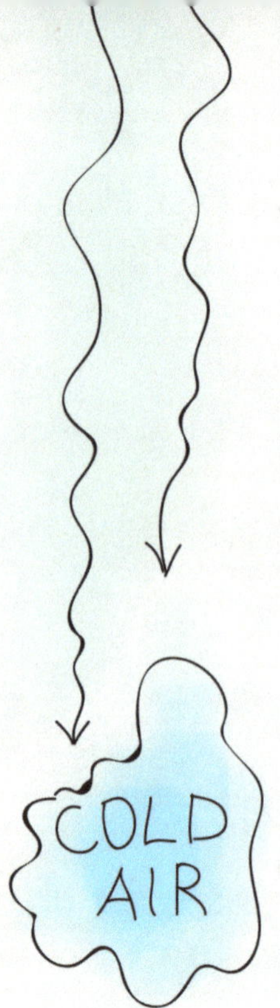

Ups and Downs of Wind

When molecules of air get cold,
they use their common sense—
they snuggle up, and that explains
why colder air is dense.

When molecules of air get warm,
much warmer than they've been,
they spread apart, and that explains
why warmer air is thin.

Now denser air is heavier
which makes the pressure high,
while thinner air is lighter and
floats up into the sky.

So put these two together and
a process will begin.
Warm air goes up, cold air comes down,
and that's what makes the wind.

Real molecules don't snuggle up,
not like we humans can,
it's just a way I think of it
to help me understand.

So when the air warms up in spring
I shout it with delight—
The temperature is changing fast,
IT'S TIME TO FLY A KITE!

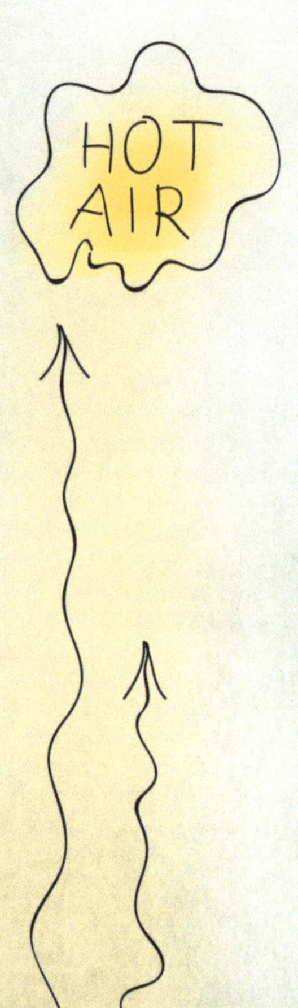

Why does a cool breeze feel so nice on a hot summer day? And why on a cold winter day does a sharp wind freeze you to the bone? The answer is **wind-chill**. Earlier we talked about the heat index, which happens when high humidity makes the air feel warmer by slowing down the evaporation of sweat from your skin. Wind-chill produces the opposite result making the temperature feel cooler than it actually is.

Heat is transferred from one body to another in one of three ways: conduction, convection, or radiation. Wind-chill is an example of convection, a transfer of heat from your body to moving air. Faster wind means more air moving across your body. More air moving across your body means a quicker transfer of heat from you to the air.

If it's a warm summer evening, you might say, "Ah...that breeze feels delightful." If it's a cold winter morning, you might say, "Ah! I'm freezing!" Or if you're the poetic type you might quote a line from the poem "Sea Fever" by John Masefield: "To the gull's way and the whale's way where the wind's like a whetted knife..." ("Sea Fever" is one of my favorite poems. I memorized it when I was in fourth grade, and I still love to recite it.)

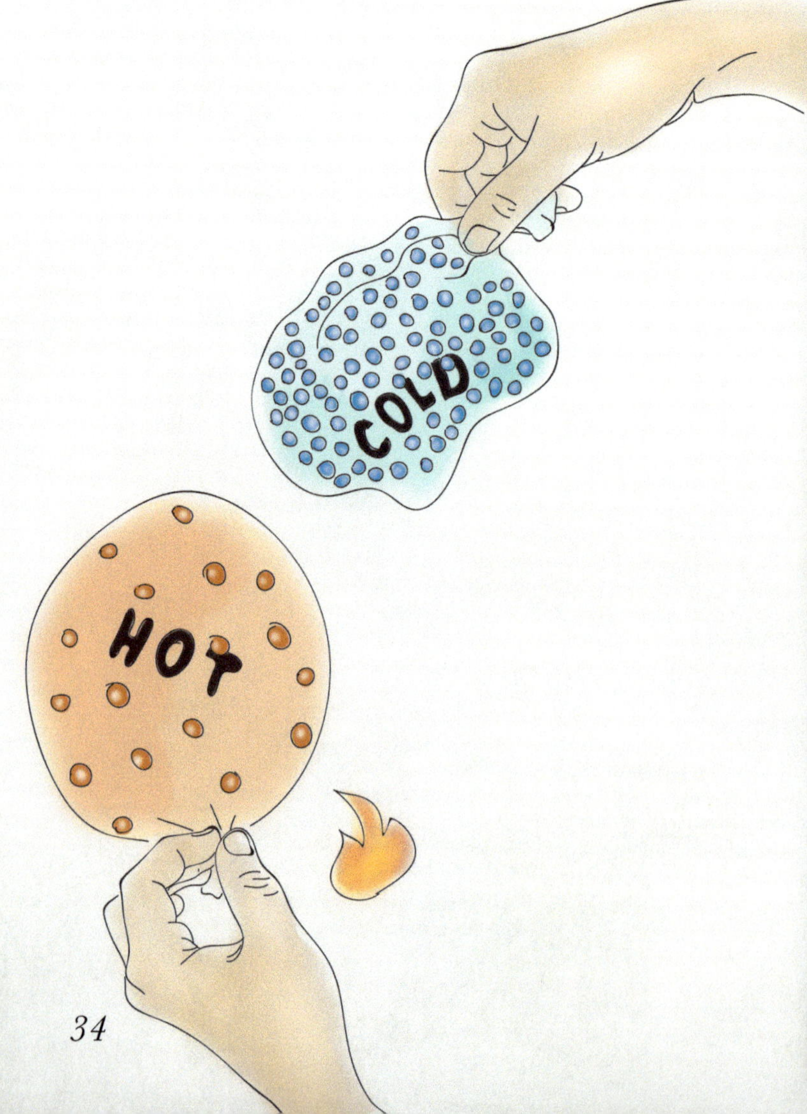
COLD
HOT

Warm-Up-Low, Cool-Down-High

At first it's a little confusing, but
it's easy once you know.
Warm air rises up and away,
which makes the pressure LOW.

Now please, don't be discouraged.
We'll give it one more try.
Cool air sinks toward the ground
which makes the pressure HIGH.

Temperature HIGH means pressure LOW.
Tricky? Yes, but now you know.
Temperature LOW means pressure HIGH.
You'll remember if you try.

This is really simple. High temperature means low pressure. This is because as air gets hotter it gets thinner. In a house, warm air rises toward the ceiling and cool air sinks toward the floor. This happens because the molecules of warm air are further apart than the molecules of cool air. Warm air is lighter and rises, cool air is heavier and sinks.

The same thing happens in the atmosphere. As this lighter warm-air starts to rise it's not pressing down so hard on the Earth's surface, which lowers the atmospheric pressure. LIGHT RISING WARM AIR = LOW PRESSURE.

Cold air is just the opposite. As air cools, it shrinks becoming denser and therefore heaver, which means it presses down harder on the Earth's surface, which increases the atmospheric pressure. HEAVY SINKING COLD AIR = HIGH PRESSURE.

How is a low pressure system like a hot air balloon? Imagine you're in the basket of a hot air balloon and you want to take off. You switch the burners on, a gas flame shoots into the opening at the bottom of the balloon, the air in the balloon gets hotter and hotter, and lighter and lighter, until it reaches the point at which it's lighter than the air outside the balloon, and it's UP-UP AND AWAY!

Kaboom!

I may be just a tiny electron,
but I can tell when something big is about to happen,
and friend, something big,
something very big is definitely about to happen.
It started just a few minutes ago.
There I was,
an electron in an H_2O molecule
forming part of a tiny crystal of ice
floating around in a gargantuan cumulonimbus cloud—
high as a mountain and dark as the floor of an ocean trench.
I was soaring in updrafts, plunging in downdrafts,
all the while crashing against other tiny ice crystals,
and the next thing I knew I was out of the ice
and in a swarm of free electrons gathered at the bottom of that cloud,
like honeybees without a queen,
like atomic orphans longing for a nuclear family,
every one of us yearning for some place to go,
which is how I know that something is about to happen,
because the first thing they taught us in Electron School
is that we free floating negative charges can't expect to stay put for long,
and now I know why,
because deep inside I feel a pull from down below,
positively charged stepped leaders reaching up to the sky,
and a bunch of us just formed into a negatively charged leader
reaching down toward the ground,
in a frenzy, craving to connect, reaching, reaching, reaching...
Oh no! Here we go!
SWOOSH! ZOOM!
FLASH! KABOOM!

Did you see that!
Our down leader met their step leader,
and suddenly we became a surge of inch thick plasma a mile long,
lighting up the sky like a new Sun,
with thunder rolling out over the land in all directions.
YES!
But hold on. What's happening now?
I'm part of a newly formed molecule.
Three oxygen atoms? Ozone!

Round and round the electron goes.
Where it stops? Nobody knows.

Here's how you can calculate the distance between you and a lightning strike. When you see the flash, start counting the seconds until you hear the thunder. Every five seconds is equal to one mile, so if it takes 10 seconds for the thunder to reach you, the strike was about two miles away. If it takes one second for the thunder to reach you, the strike was in the next block.

This is how it works. Light travels at a speed of 186,000 miles per second, so you will see the flash of a lightning strike at almost the same moment it happens. Sound travels much more slowly, only about 760 miles per hour at sea level, or about one mile every five seconds. By counting the seconds between the flash and the thunder, you can calculate the distance between you and the strike.

All in the Family
An Interview with Earth's Atmospheric Gases

(A Play In One Act)

Cast of Characters:

Talk show host: Harold Hotair (Stuffy and a little pompous.)
Main Guest: Oralinda Oxygen, known as O_2 (Husky voice, femme fatale.)
Other Guests: Nicky Nitrogen
Reginald Argon
Clair Carbon Dioxide
Olivia Ozone
Wally Water Vapor

(Harold Hotair, host of TV talk-show, The Atmosphere: It's a Gas, is seated at a desk conducting an interview with Oralinda Oxygen.)

Harold:

Good evening, ladies and gentlemen,
this is your host Harold Hotair
and welcome to another edition of *The Atmosphere: It's a Gas.*
Tonight's guest is someone none of us could live without.
With the symbol O and the atomic number 8,
she leads the chalcogen group of the periodic table,
highly reactive, she forms oxides with most other elements,
the third most abundant element in the universe
and a full one fifth of Earth's atmosphere,
the flame in the generator of every human cell,
won't you please help me welcome...
the one, the only, Ms. Oralinda Oxygen!

(Audience applauds, host turns to his guest)

Harold Hotair:

Hello, Ms. Oxygen,
and welcome to *The Atmosphere: It's a Gas.*
And may I call you Oralinda?

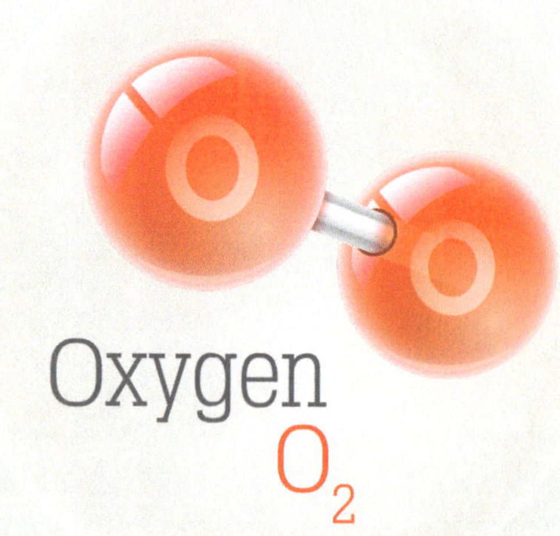

Oralinda Oxygen:
(With a flirty smile)

You may call me Oralinda if you wish,
but those who know me best refer to me as O_2.

Harold Hotair:

Then I'll call you O, too.
Now what I would...

Oralinda Oxygen:

No darling.
Not O-too as in also.

(Laughing mischievously, and drawing in the air the letter "O" and the number "2.")

O...2...like the number,
which means my molecule consists of
two bonded oxygen atoms.
And thank you for inviting me on your show.
You know how I love to react with a live audience.

Harold:

Yes, you are indeed reactive.
If I'm not mistaken,
half of Earth's crust consists of chemical compounds and oxides
that are the result of your legendary reactivity.

Oralinda Oxygen:

Now daaarling,
Why are we talking about dirt?
I thought we were here to talk about air.

Harold Hotair:

(As Harold introduces the other atmospheric gases, they each come center stage, bow to the studio audience as the audience applauds, and takes a seat across from O_2.)

Indeed we are,
which brings me to tonight's special surprise.
As all of our viewers know,
the air of Earth's atmosphere is not just one gas
but a mixture of several gases,
and how can we talk about air without the others.
So let me introduce...
Nicky Nitrogen.
...
Reginald Argon.
...
Claire Carbon Dioxide.
...
Little Olivia Ozone.
...
And last but not least, Mr. Muggy Day himself
Wally Water Vapor.
Welcome to all of you
and thank you for joining us on *The Atmosphere: It's a Gas.*

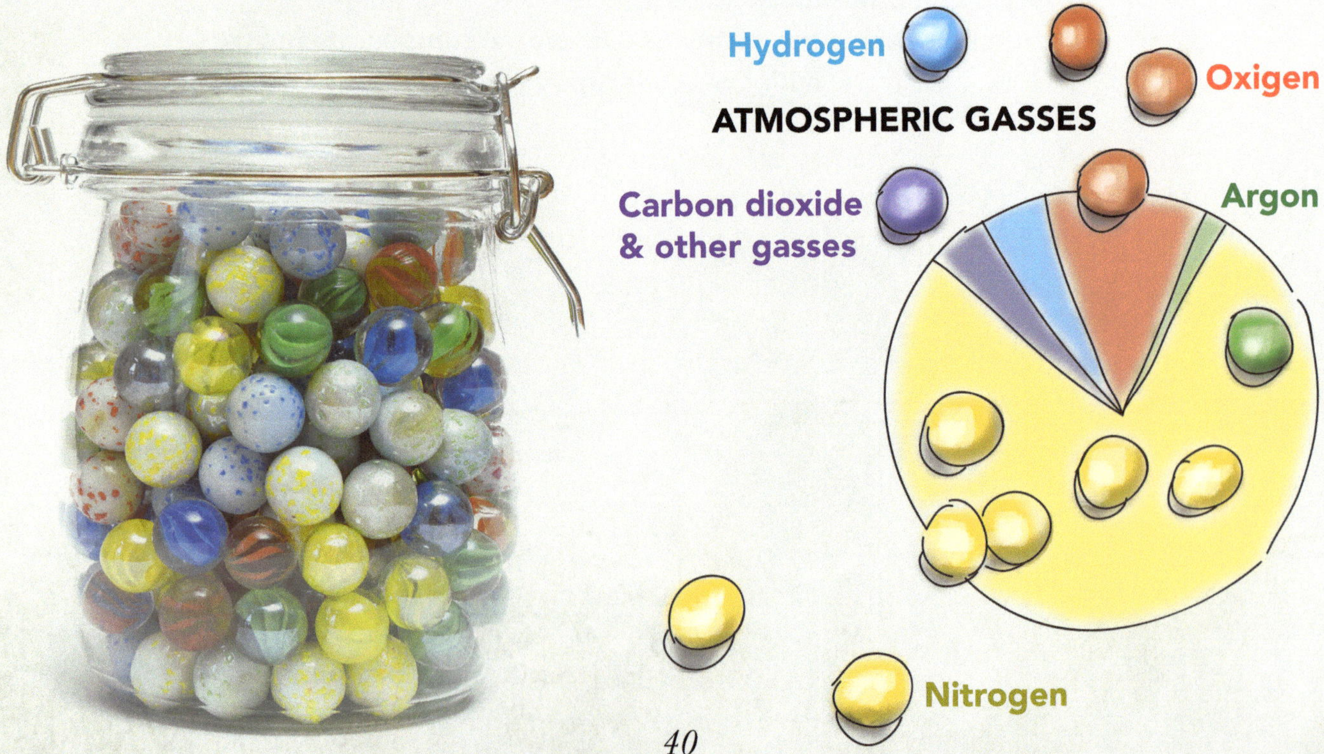

Oralinda Oxygen:

Well, Harold.
I must say, I feel a bit ambushed.
Sure, I'm not the only gas in the atmosphere,
but I may as well be.
After all, when someone says, I need some air,
what they really need is Oxygen.
All these...others are just fillers.

Nicky Nitrogen:

Filler! Filler?
Nobody calls nitrogen filler.
I make up 78% of the of Earth's atmosphere,
which means that if the atmosphere were a jar of a hundred marbles,
78 of those marbles would be me.

Oralinda Oxygen:

Oh dear. There you go again, Nicky,
bragging about your majority position.
Well the atmosphere is not a democracy,
it's not about numbers,
it's about popularity,
and I am obviously the most popular puff of atmospheric stuff.
And besides, nitrogen comes from a Greek word for choke.
Need I say more?
And what are you shaking your head about, Mr. Reginald Argon.
You're less than 1% of the atmosphere.
Less than one little marble out of a hundred.
And you interact with nothing,
so much so that your name comes from the Greek word for lazy.

Reginald Argon:

Yes, that's one translation,
but inactive would be more accurate,
and the scientific word for inactive is inert.
I am one of the noble gases,
which means that I am colorless and odorless,
my molecules consist of one atom,
and I have very low chemical reactivity,
which is a good thing.
Why Oralinda, if the atmosphere were a hundred percent oxygen,
most everything on Earth would start burning
and wouldn't stop till the world was a pile of ashes.

Oralinda Oxygen:

Why, thank you, Mr. Noble Gas Argon.
I have a reputation for being HOT,
but it's nice to know you noticed.

Olivia Ozone:

Don't stoop to her level, Reggie.

(Turning toward Oralinda Oxygen)

I may be your sister—
you're O_2 and I'm O_3,
but that extra atom makes a world of difference.
And before you make fun of my name,
I know it comes from the Greek word for smell,
and I do have a strong odor,
but I'm important because of what I do,
not how I smell,
and what I do is protect life on Earth.
I filter those destructive wavelengths of ultraviolet light.

(Beginning to get a little weepy)

And I know I'm not very popular.
It was just a few years ago
my little ozone layer was on its way to being totally depleted
by those wicked CFCs,
those chloro-fluoro-carbons humans used in air conditioners
and deodorant, of all things,
but they woke up before it was too late,
and...and...

(Breaking out into tears)

Why do you have to be so mean?!

Oralinda Oxygen:

I'm sorry, little sister.
Please don't cry.
If it makes you feel any better,
my name comes from the Greek word for sharp as in sour taste,
because that Frenchman who discovered me
made a mistake about me and acids,
but it's no mistake that, at times, I can be a little sharp.

(Turning toward Claire Carbon Dioxide)

And what about you, scarecrow.
What are you gawking at, Ms. Claire Carbon Dioxide.

Clair Carbon Dioxide:

Hmmm! I was just wondering when you'd be coming after me.
Figured you'd be blaming me for all this climate change—
global temperature rising, ice caps melting,
sea levels higher and higher,
shifts in ocean currents
causing humongous changes in weather patterns,
and who knows what else.
Everybody knows I'm a greenhouse gas,
and that all this climate change is a greenhouse thing.
And if the truth be known, it is me,
with a little help from Wally Water Vapor over there,
and our little buddy Mervin Methane.
Where is he anyway?
Probably too busy watching all those cattle and sheep and goats,
every one of those little grass grazers
belching and flatuating all that methane gas,
but nobody ever talks about ole Mervin Methane,
it's always me, big-bad Ms. CO_2.
So I just figured I was next on your list to make fun of.

Oralinda Oxygen:

I must seem terrible to all of you, and I guess I deserve it.
Yes Claire, you have protected Earth for a long time,
you and those other greenhouse gasses like a blanket
holding in some of that long-wavelength infrared-radiation
keeping it from escaping into space
so Earth doesn't freeze over like naked toes on the Yeti.

(Turning toward the audience)

Sorry. I know that's a little out of character,
but sometimes I just can't help myself.

(Turning back to Claire Carbon Dioxide)

And most everybody knows that our greenhouse blanket has gotten too thick,
and that it's changing Earth's climate pretty drastically.

(Turning back toward the audience)

And most scientists agree that humans are causing it
with all those factories and cars and trucks and cows and goats,
like a big smoke stack pumping out the gases that used to be their heroes
but are now like the four horsemen of the Apocalypse coming after them.

(She takes a breath and softens her voice)

Look, I don't want to be playing the blame game,
it's pretty clear how all this happened,
and it all happened so quickly—
some good science, harnessed electricity, internal combustion engines,
oil and gas giving everybody power,
making lots of people rich,
everybody loving it,
until a bunch of scientists come along and spoil the party.
So now it's time to change.
And we gases all around you in the atmosphere,
we like you humans,
we hope you stick around for a long time,
and we believe in you.
It's not too late.
You can do it.
Why, who knows what new technologies you'll come up with.
Probably make you richer and happier than ever.

(Turning to Wally Water Vapor)

Well Mr. Wally Water Vapor,
you've been awfully quiet over there.
Why don't you tell us what's on your mind,
what do you have to say for yourself?

Wally Water Vapor:
(Hesitates, scratches his head, looks all around with a blank stare, and finally turns toward Oralinda Oxygen)

Well, Oralinda.
At first I was thinking that maybe I should form myself into some water droplets
and precipitate a rain shower to cool off our little gathering,
but things seemed to have chilled out,
so I'm cool too.
Actually, I'm hot and muggy, but for me that's cool.

Harold Hotair:

Well, that's it for this week's edition of *The Atmosphere: It's a Gas.*
Please join us next week
when our guest will be that world famous decorator of the northern sky,
Madam Aurora Borealis,
who will reveal the magical secret of converting solar wind
into shimmering curtains of atmospheric color.

(Audience applauds. The End.)

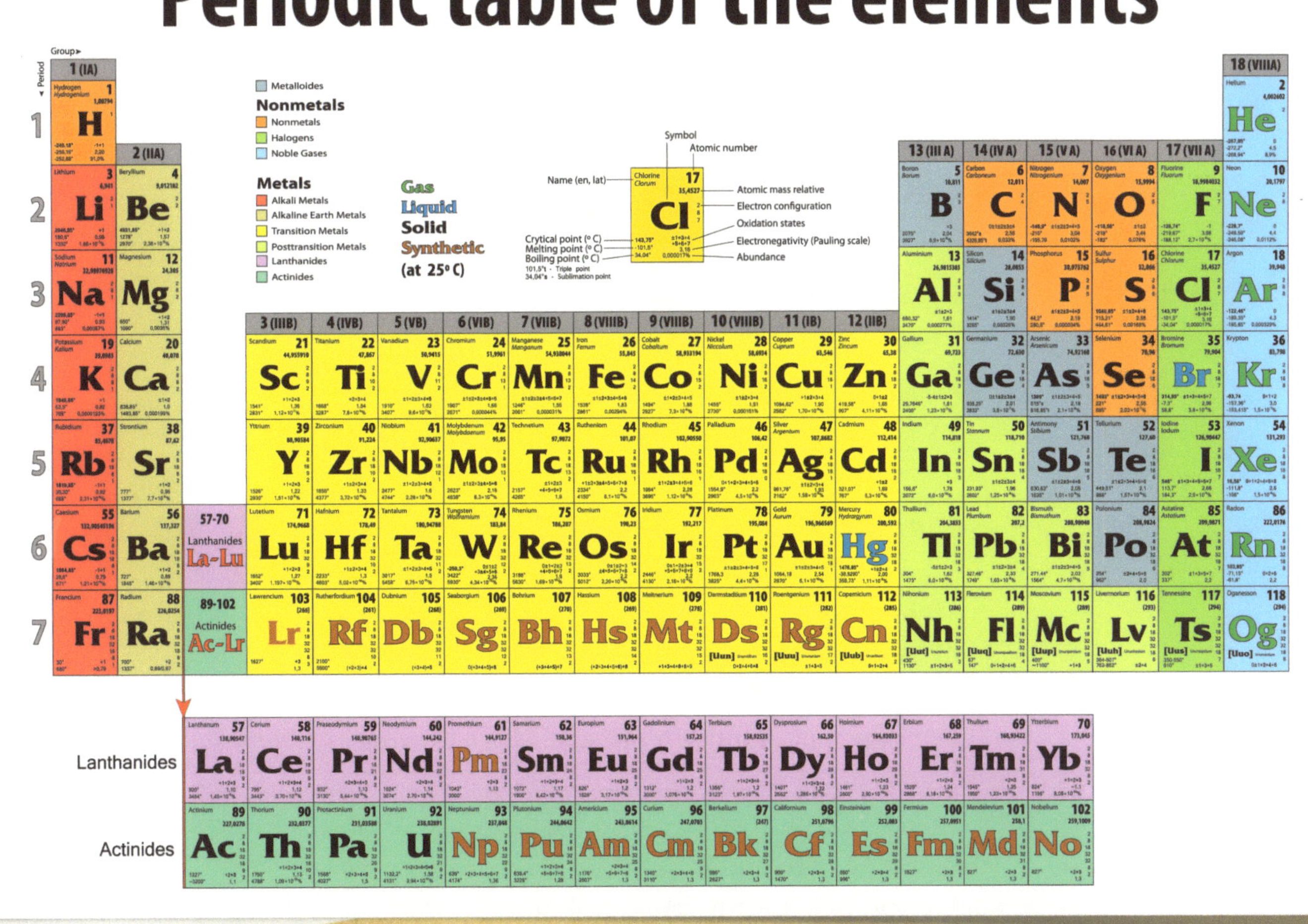

Climate change, or global warming, is presently the greatest danger to life on Earth. Earth's five previous mass extinctions were all caused by events that altered the composition of Earth's atmosphere resulting in deadly changes in global climate. The first four were caused by prolonged volcanic eruptions. The cause of the fifth, the one that caused the extinction of the dinosaurs, is still a matter of scientific debate. Scientists think it was caused by either a major asteroid strike, or by massive volcanic eruptions in India, or by both. In either case, the result was an alteration in the composition of Earth's atmosphere with the same deadly results.

So why is the atmosphere so important in regulating climate? The answer is very simple. It's all about how much of the Sun's energy the atmosphere lets in, and how much of that energy the atmosphere releases back into outer space. The result is a balance that controls Earth's climate. If the balance is upset, our climate changes.

Once again, Earth is threatened by the prospect of climate change resulting from changes in Earth's atmosphere. The overwhelming consensus among scientists is that our current climate change is being caused by a dramatic increase in greenhouse gases produced by human beings. As our atmosphere fills with more and more greenhouse gases, it holds in more and more of the Sun's energy, and Earth's climate becomes warmer and warmer resulting in a long list of catastrophic consequences.

Greenhouse Gas Chamber
or
End the Big Burp

CO_2! CH_4!
Enough of them, we need no more.
CH_4! CO_2!
Hurting me and hurting you.

Life isn't always easy,
and it isn't always nice,
and lots of things we love to do
come with a hidden price.

Like gasoline in cars and trucks,
we need to end it now.
And this one's pretty easy 'cause
already we know how.

This other one is harder,
but we've got to end it too—
a billion belching cattle
making methane with their poo.

CO_2! CH_4!
Enough of them, we need no more.
CH_4! CO_2!
Hurting me and hurting you.

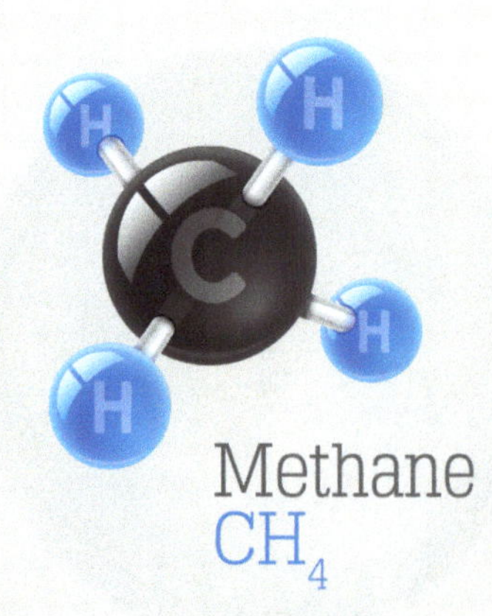

We need our greenhouse gases
to protect our inner space,
but adding more is making earth
a very dangerous place.

CO_2 is the chemical formula for a gas called **carbon dioxide**. It contains one carbon molecule bonded with two oxygen molecules. CH_4 is the chemical formula for a gas called **methane**. It contains one carbon molecule bonded with four hydrogen molecules.

Excess amounts of carbon dioxide and methane in Earth's atmosphere is the principle cause of the gradual increase in global temperature and the engine of what scientists call climate change.

Like a blanket holding heat
but not like warming hands and feet.
It's heating up the atmosphere,
getting hotter every year.

Melting ice, rising sea,
much less land for you and me.
Making Earth the perfect place
to toast our little human race.

So listen very closely.
These words are really true.
CO_2!...CH_4!
Enough of them, we need no more.
CH_4!...CO_2!
What happens next to us on Earth?
It's really...up...to you!

In My Hands

Planet Earth's my home.
I like it here a lot.
Perfect? Not at all, but it's
the only home I've got.

And our climate is a system,
a planetary scheme—
sunlight, water, atmosphere
all working as a team.

It's a system we depend on
so we've got to put it first,
but scientists are saying that
it's changing for the worse.

That's why I need to understand
our climate through and through
so I can help to figure out
exactly what to do.

'Cause planet Earth's my home.
I like it here a lot.
Perfect? Not at all. But it's
the only home I've got.

More than 97% of the scientists who study climate change have concluded that it is really happening, that human-made greenhouse gases are causing it, that there is something we can do about it, and that we have to start immediately.

It's the overwhelming consensus of the world scientific community, and there is no reason to distrust it. This is the same scientific community that has eliminated deadly diseases, put us in outer space, and created the digital world of computers. Sure, nobody's perfect, and even scientists make mistakes, but not 97% of them at the same time.

So what do you think? Should we stand by and do nothing as greenhouse gas concentrations bring us ever closer to predictable disaster? Or is it time to take a stand? It really is in your hands.

An Invitation

Some years ago it became my life's dream to write a bunch of poems about science; poems that would be both fun and instructional; poems to simultaneously entertain you and help you become lifelong learners.

I've been at it now for over 20 years, and I have begun to understand that this little book and the ones to follow, are just a first step, and that the next step depends on you.

That's why I am inviting you to join me in the wonderful work of awakening the joy of powerful thinking in the minds of your generation. We humans are proud of our tall buildings and our 2,000 year old aqueducts and our 4,000 year old pyramids, and we should be. But what we can be most proud of is that pocketful of simple yet powerful ideas that made every one of those things possible. I invite you to join me in helping to keep those ideas healthy and strong.

Look back through the pages of this book, pick out one of the pieces that you really like, and then do something exciting with it. Send it to my email at **brod@brodbagertsheartofscience.com**. You might make a video of you and your friends performing it; or create your own original illustration or animation; or write your own original poem or play; or simply write your thoughts after reading it; or maybe something else I can't even imagine. As time goes by we at **Brod Bagert's HeART of Science** will be selecting some of your material and, with your permission, share it with others just like you.

I believe that many of you will accept this invitation and have fun doing it. There may even be a few of you who will like it a lot, a few of you for whom it will become the joy of your lives, as it has been in mine. Maybe you'll be the ones to help take **Brod Bagert's HeART of Science** to a whole new level. And maybe, eventually, you'll become the stars that bring light to the night skies of your generation. Please accept this as your invitation.

www.ingramcontent.com/pod-product-compliance
Lightning Source LLC
LaVergne TN
LVHW071202160826
845679LV00003B/720

* 9 7 8 1 7 3 2 1 5 1 5 4 3 *